OnBoard
ACADEMICS

Addition and Subtraction

© 2015 OnBoard Academics, Inc
Portsmouth, NH
800-596-3175
www.onboardacademics.com
ISBN: 978-1-63096-067-4

OnBoard Academic's books are specifically designed to be used as printed workbooks or as on-screen instruction. Each page offers focused exercises and students quickly master topics with enough proficiency to move on to the next level.

OnBoard Academic's lessons are used in over 25,000 classrooms to rave reviews. Our lessons are aligned to the most recent governmental standards and are updated from time to time as standards change. Correlation documents are located on our website. Our lessons are created, edited and evaluated by educators to ensure top quality and real life success.

Interactive lessons for digital whiteboards, mobile devices, and PCs are available at www.onboardacademics.com. These interactive lessons make great additions to our books.

You can always reach us at customerservice@onboardacademics.com.

Addition & Subtraction

Key Vocabulary

addend

sum

difference

number sentence

   www.onboardacademics.com

Give Matteus the same number of strawberries as Alison.

Hint; first count Alison's strawberries and that will help you to decide how many to add to Matteus.

$$4 + \boxed{} = \boxed{}$$

_____

Addition counting on

$$5 + 3 = \boxed{?}$$

Look at the illustrations.

Start with five and count three more; 6, 7, 8. to get the answer.

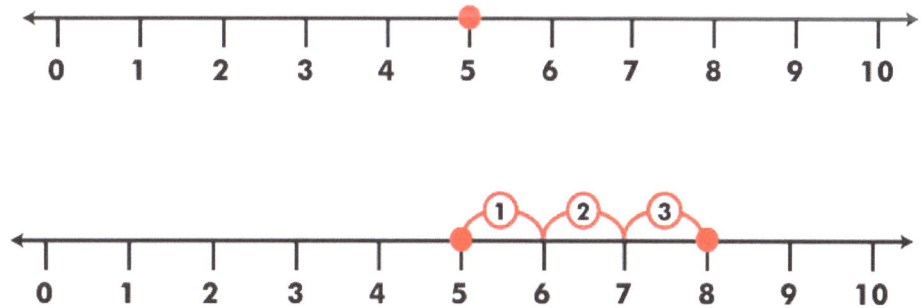

Practice Addition.
Use the number line to help count on.

2 + 5 =

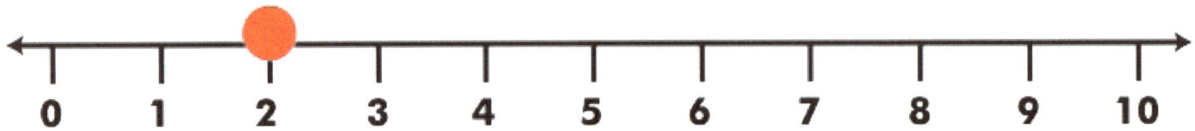

6 + 3 =

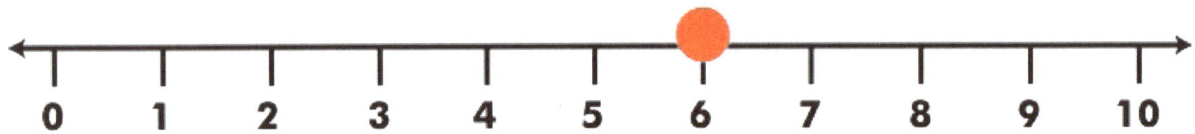

5 + 2 =

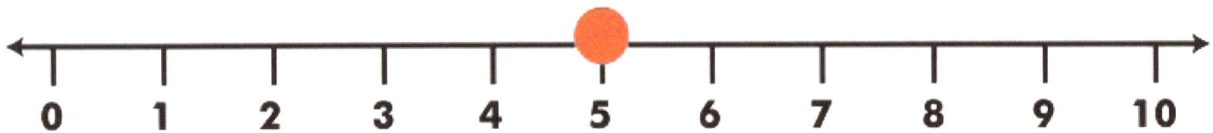

3 + 6 =

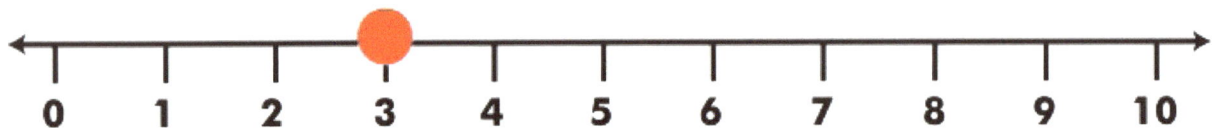

www.onboardacademics.com

Counting back
Look at the illustrations.

Start with eight and count back three; 7, 6, 5, to get the answer.

$$8 - 3 =$$

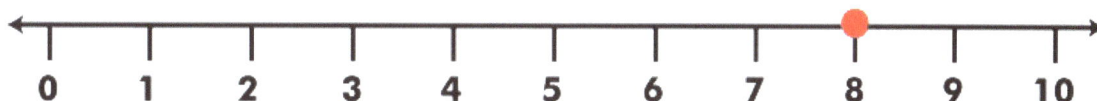

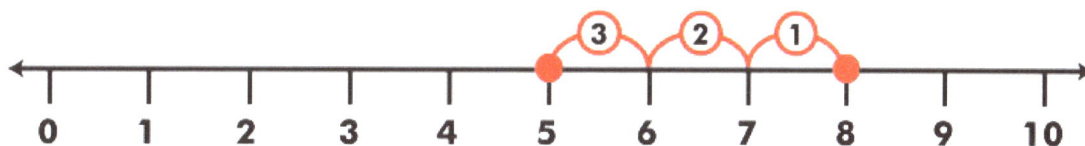

_____

Practice counting back

$$10 - 4 =$$

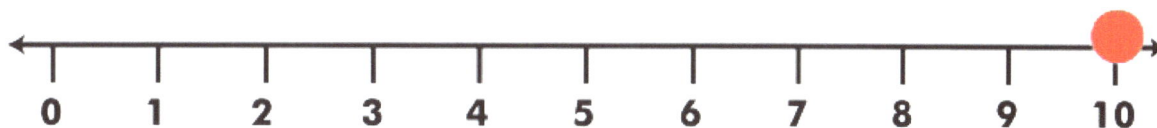

$$5 - 5 =$$

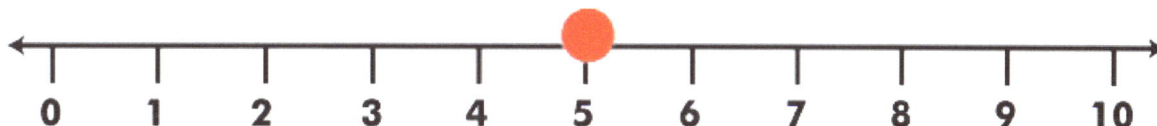

Practice addition and subtraction

1   1 + 4 = ☐

2   5 − 1 = ☐

3   7 + 2 = ☐

4   9 − 2 = ☐

5   6 + 4 = ☐

6   10 − 6 = ☐

————————

The 4, 5, 9, fact family.

Look at the equations you can make with these three numbers. They are related.

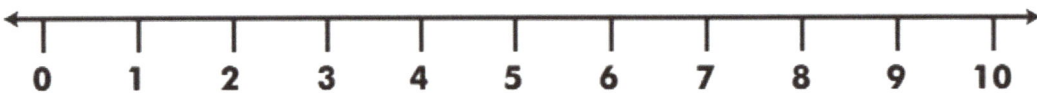

```
0   1   2   3   4   5   6   7   8   9   10
```

4   **5**   **9**

5 + 4 = ☐   4 + ☐ = 9

9 − ☐ = 5   9 − ☐ = 4

Stretch Exercise
Solve these word problems. Write the equation and the answer.

1 Tori has 3 erasers. Owen has 4 erasers.
How many erasers are there altogether?

2 James has 10 cookies. He eats 4 cookies.
How many cookies does he have left?

3 Alicia has $9. She spends $7 on a gift for her sister and
$2 on a soda. How much money does she have left?

Name_____

Addition and Subtraction Quiz

Circle or fill in the correct answer.

1   **True or false? 11 − 3 = 8**

2   **Which problem equals 12?**

     A   **4 + 5 = ?**

     B   **7 + 5 = ?**

     C   **3 + 8 = ?**

     D   **5 + 6 = ?**

3   **9 − 4 = ?**

4   **What is the missing value?   6 + ? = 9**

Addition and Subtraction

Key Vocabulary

Sum

Difference

Addend

Sort the sums.
Quick add the problems and place the sums in the proper boxes.

| 18 + 30 | 87 + 53 | 27 + 20 | 55 + 58 |

| 78 + 43 | 14 + 23 | 48 + 89 | 49 + 52 |

More Than 100	Less Than 100

Addition using an algorithm.

An Algorithm is a step-by-step solution to a problem. It is the rule for completing the problem.

Here's an example: an algorithm for adding two digit numbers is to add the ones and then the tens and then combine the answers.

Does it matter if we change the order?

tens ones

$$\begin{array}{r} 3\ 7 \\ +\ 5\ 5 \\ \hline \\ \hline \end{array}$$

Addition using partial products.

Here we add the tens first, the blue numbers. After that, we add the ones, the brown numbers.

Once we have the totals of the tens and the ones, we add them together to get a grand total.

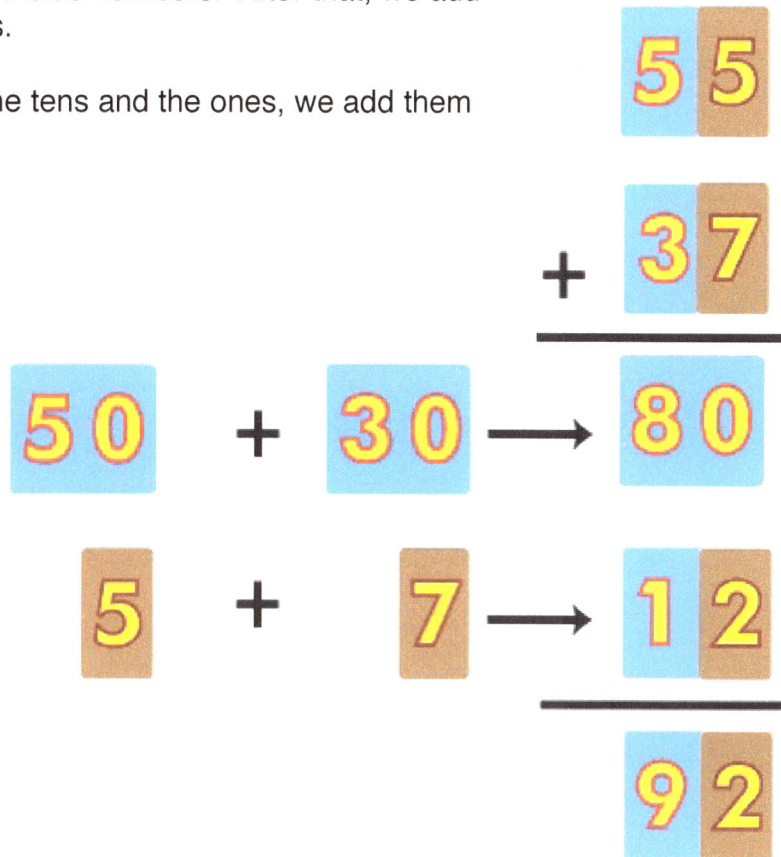

Subtraction using an algorithm

Does it matter if we change the order?

```
  2 4 3
-   6 8
_____


_____
```

Subtraction by counting up. That's weird.

$$244 - 63 = \boxed{?}$$

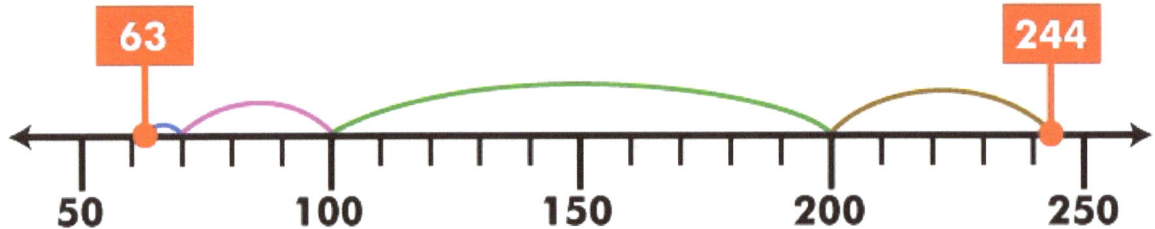

$$63 + 7 = 70$$

$$70 + 30 = 100$$

$$100 + 100 = 200$$

$$200 + 44 = 244$$

$$7 + 30 + 100 + 44 = 181$$

Start with the smaller number (63 in this case) and count up to a friendly number by adding 7 to get 70. Then we add 30 to get another friendly number, 100 is easy to add to that to get 200. Then we add the final amount to get to our highest number in the equation (244 in this case).

Now its easy to add up the friendly numbers to find the difference between 244 and 63.

Practice addition and subtraction.

1

```
  1 8 5
+   7 4
_____


_____
```

3

```
  6 7 4
+ 3 5 8
_____


_____
```

3

```
  5 9 3
-   6 7
_____


_____
```

4

```
    9 2
-   1 9
_____


_____
```

Solve these addition and subtraction problems.

1) $33 + \boxed{} = 152$

2) $97 - \boxed{} = 23$ **Subtract the solution.**

3) $\boxed{} - 33 = 56$ **Use addition to solve.**

4) $553 + \boxed{} = 660$

5) $782 - \boxed{} = 535$

6) $114 - \boxed{} = 67$

Here are the answers but not in the correct order :)

247   119   47

74   89   107

Practice with the Carvahlo brothers.

65 inches
56 inches
47 inches

Jesús   Matteus   Leonardo

1) Jesús is how much shorter than Leonardo?

2) Leonardo is how much taller than Matteus

3) Matteus is how much shorter than Jesús?

Name_____

Addition and Subtraction Quiz

Fill in or circle the correct answer.

1 Changing the order of the numbers in a subtraction problem will not change the answer.

2 Which problem will equal 78?

- **A** 54 + 33
- **B** 54 + 23
- **C** 54 + 25
- **D** 54 + 24

3 546 + 87 = ?

4 James has 231 baseball cards. He gives 54 to his younger brother. How many cards does he have left?

Place Value

Key Vocabulary

Digit

Ones

Tens

Hundreds

What value is represented by each of these blocks? Write the answer in the box.

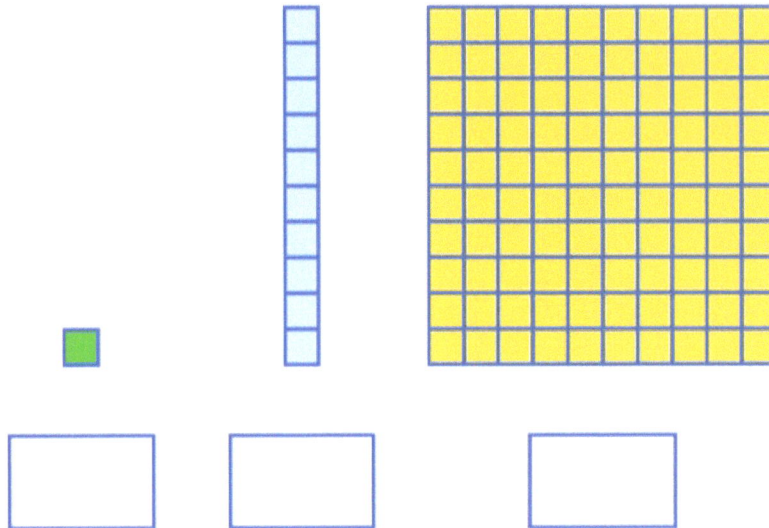

What number do the blocks represent? Hint; count the number of rows of ten blocks and then the number of single blocks (green).

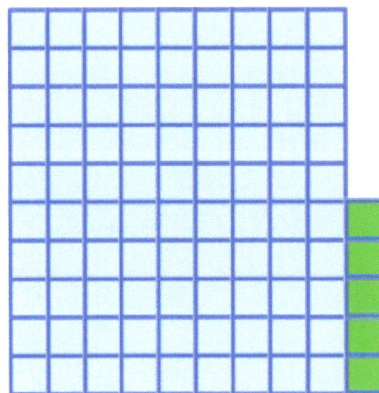

	+		=	
tens		**ones**		**Answer**

What number do these blocks represent?

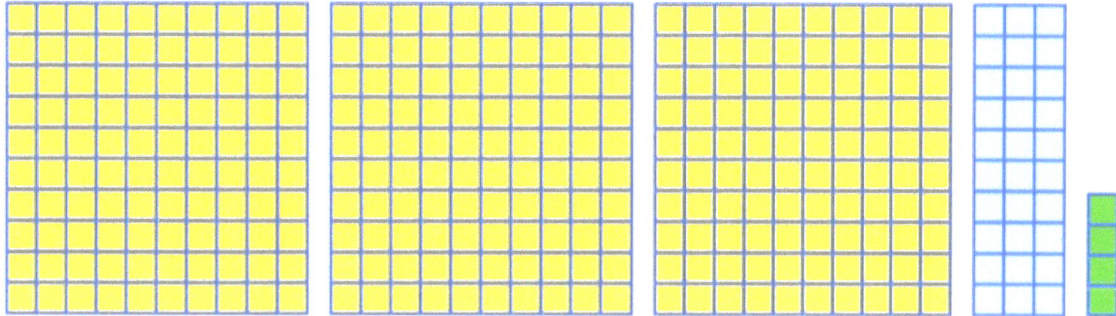

| hundreds | + | tens | + | ones | = | Answer |

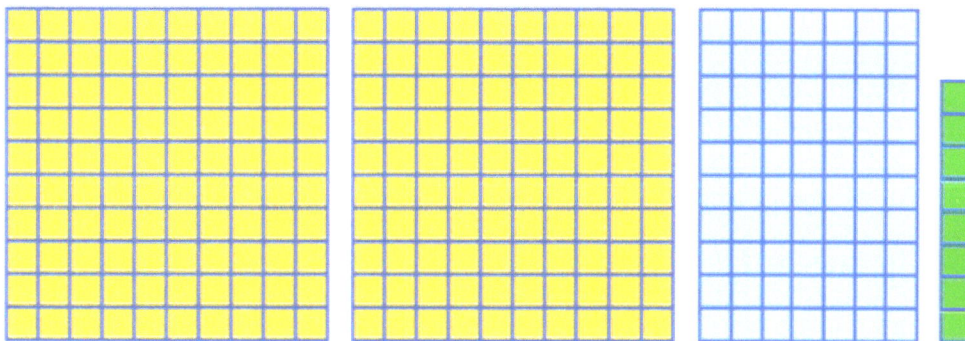

| hundreds | + | tens | + | ones | = | Answer |

How many of each of the 100 blocks, ten blocks and ones blocks do you need to make these numbers? Write the answer in the space provided.

	hundreds +		tens +		ones =	**230**

	hundreds +		tens +		ones =	**333**

What is the value of the red 5? The first one is done for you.

hundreds	tens	ones

3 5 7

50

2 7 5

5 7 0

Name_____

Place Value Quiz

Circle or fill in the correct answer.

1   **8 6 5** = 80 + 60 + 5

2   **What number do the blocks in Figure 1 represent?**

   A 66
   B 68
   C 58
   D 108

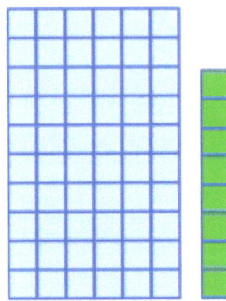

Figure 1

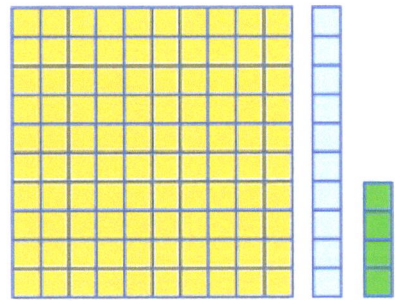

Figure 2

3   **What number do the blocks in Figure 2 represent?**

4   714 = 700 + ____ + 4

Place Value

Key Vocabulary

Less than

Greater than

Equal to

Compare

Order

Which Island is the hottest?

Draw the sun over the hottest island.

Santa de Sizzla   **Isle de Scorchio**

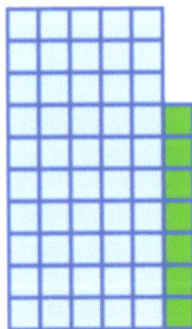

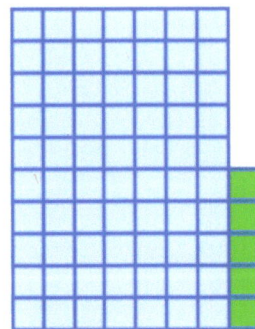

5 tens + 7 ones   **7 tens + 5 ones**

Which product is the Star Buy?

The Star Buy is the best or lowest price. Draw a star in the product's circle to indicate the Star Buy.

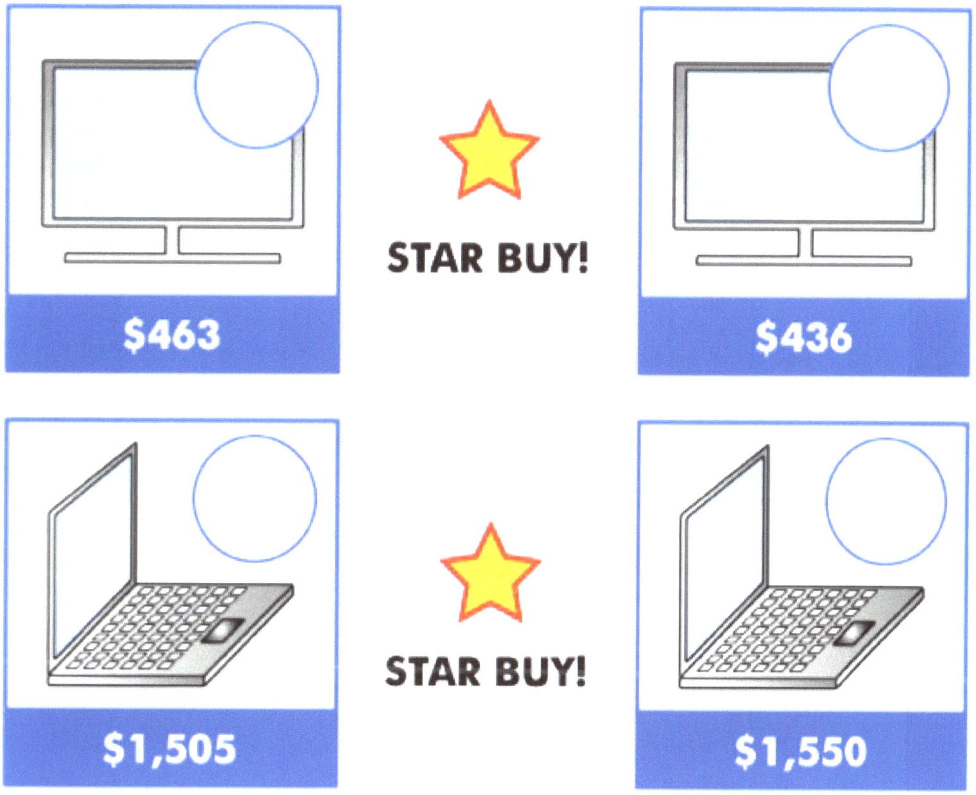

$463

STAR BUY!

$436

$1,505

STAR BUY!

$1,550

Investigate place value.

Fill in the boxes below

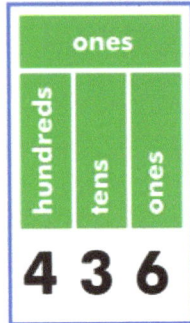

ones		
hundreds	tens	ones
4	3	6

☐ hundreds

+ ☐ tens

+ ☐ ones

ones		
hundreds	tens	ones
4	6	3

☐ hundreds

+ ☐ tens

+ ☐ ones

Use the symbols < > to compare the weight of Ella and Eli.

Write the weight of the smallest elephant in the box above to identify ones, tens, hundreds and thousands.

Ella		Eli
4,5 4 2 lb	○	**4,4 5 2 lb**

thousands			ones		
hundreds	tens	ones	hundreds	tens	ones

< less than

> greater than

_____

Investigate place value with thousands.

1,5 0 5   1,5 5 0

☐	☐	thousands
+ ☐	+ ☐	hundreds
+ ☐	+ ☐	tens
+ ☐	+ ☐	ones

thousands			ones		
hundreds	tens	ones	hundreds	tens	ones

Do you get < and > mixed up.
Use these alligators giant jaws as a reminder. The alligator always eats the bigger number.

greater than

452 > 341

341 < 452

less than

"I always eat the biggest number"

What are the largest numbers and the smallest numbers that you can make with the digits 3, 5, and 9.

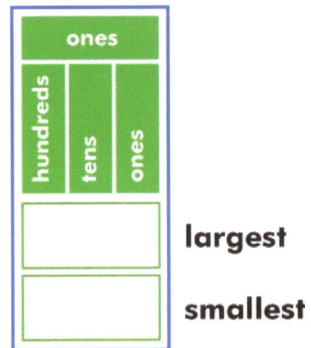

ones		
hundreds	tens	ones
			largest
			smallest

3 5 9

Fill in the correct symbol.

1	689	986	**2**	24,834	32,834
3	1,324	1,234	**4**	65,234	65,434
5	11,324	11,624	**6**	74,234	77,004

(<) (>)

1	45	54	**2**	89	78
3	101	110	**4**	735	753
5	463	436	**6**	989	898

(<) (>)

Name_____

Place Value Quiz

Circle or fill in the correct answer.

1 **True or false? $945 < $954**

2 **Write 9 hundreds + 5 tens + 0 ones in standard form.**
- **A** **905**
- **B** **950**
- **C** **509**
- **D** **590**

3 **Write 5 thousands + 0 hundreds + 0 tens + 3 ones in standard form.**

4 **Write 9 thousands + 9 hundreds + 4 tens + 7 ones in standard form.**

www.ingramcontent.com/pod-product-compliance
Lightning Source LLC
Chambersburg PA
CBHW052045190326
41520CB00002BA/196